INVASIVE SPECIES

BIGHEADED CARP

by Martha London

FOCUS READERS

NAVIGATOR

WWW.FOCUSREADERS.COM

Focus Readers is distributed by North Star Editions:
sales@northstareditions.com | 888-417-0195

Produced for Focus Readers by Red Line Editorial.

Content Consultant: Dr. Peter W. Sorensen, Professor of Fisheries, Wildlife, and Conservation Biology at the University of Minnesota

Photographs ©: Ryan Hagerty/USFWS, cover, 1; Shutterstock Images, 4–5, 8–9, 11, 18 (silver carp), 18 (bighead carp), 18 (child), 21; Jon Amberg/USGS, 7; Red Line Editorial, 13; Nerissa Michaels/Illinois River Biological Station/Detroit Free Press/AP Images, 15; USGS, 16–17, 22–23, 25; Jessica Vandrick/US Army, 27; Lee Roberts/US Army Corp of Engineers, 29

Library of Congress Cataloging-in-Publication Data
Names: London, Martha, author.
Title: Bigheaded carp / Martha London.
Description: Lake Elmo, MN : Focus Readers, [2022] | Series: Invasive species | Includes index. | Audience: Grades 4-6
Identifiers: LCCN 2021003729 (print) | LCCN 2021003730 (ebook) | ISBN 9781644938539 (hardcover) | ISBN 9781644938997 (paperback) | ISBN 9781644939451 (ebook) | ISBN 9781644939864 (pdf)
Subjects: LCSH: Bighead carp--Juvenile literature. | Introduced fishes--Juvenile literature. | Pest introduction--Juvenile literature. | Nature--Effect of human beings on--Juvenile literature.
Classification: LCC QL638.C94 L66 2022 (print) | LCC QL638.C94 (ebook) | DDC 597/.482--dc23
LC record available at https://lccn.loc.gov/2021003729
LC ebook record available at https://lccn.loc.gov/2021003730

Printed in the United States of America
Mankato, MN
082021

ABOUT THE AUTHOR

Martha London is a writer and educator in Minnesota.

TABLE OF CONTENTS

CHAPTER 1

SEARCHING FOR FOOD

A buffalo fish swims along the bottom of the Mississippi River. The fish is looking for tiny organisms to eat. But it cannot find enough food. That's because the Mississippi River has become crowded with silver carp. This **species** of bigheaded carp eats similar foods as the buffalo fish. With so many carp, there is

Young silver carp tend to be more silvery than adults.

less food for the other fish in the river. This puts the buffalo fish in danger.

There are two kinds of bigheaded carp in North America. Silver carp is one. The other is called bighead carp. Its name can be confusing because it is very similar to the group's name. But bighead carp is a specific species of carp. And bigheaded carp (ending with "-ed") is the group.

Bighead and silver carp are **filter feeders**. They eat **algae** and other tiny organisms. They also eat tiny particles and bits of dead plants and animals.

Both fish are **invasive species** in the United States. People brought the two species there in the 1970s. The carp

A scientist holds a silver carp (left) and a bighead carp.

spread into rivers draining into the Mississippi. Huge numbers of carp now live in many rivers of the central United States. They have changed aquatic **food webs** in a number of ways.

CHAPTER 2

COMING TO AMERICA

Bigheaded carp live in freshwater rivers and lakes. Both species are **native** to parts of China. Silver carp are also native to eastern Russia and Vietnam. In native areas, the numbers of bigheaded carp are shrinking. There are a few reasons why. One reason is dams. People have built dams in many rivers. Dams block

The Three Gorges Dam on the Yangtze River in China has lowered native populations of bigheaded carp.

carp from swimming to certain parts of the rivers. So, carp cannot breed in their usual places. When they do breed, their young may die.

Another reason is overfishing. For instance, silver carp is a food source in parts of China and Russia. But people often catch too many silver carp. A third reason is water pollution. Carp struggle to survive in polluted water.

In the 1960s and 1970s, fish farmers and scientists brought several carp species to Arkansas. They were working with government officials. Experts incorrectly thought there was little risk. Farmers placed carp in farming ponds.

The algae that bigheaded carp eat is often green in color.

The fish ate algae in the ponds. They kept the ponds clean. Scientists brought in the fish to study. People also began selling carp at fish markets.

Over time, some bigheaded carp escaped. People discovered them in the wild in the 1980s. The fish began spreading through the Mississippi River **basin**. This river system contains the bodies of water that flow into the Mississippi River. Water from Canada and dozens of US states forms the basin.

As a result, bigheaded carp were able to spread great distances. By 2020, they were thriving across the central United States. They had reached the southern

KINDS OF CARP

Silver and bighead carp are just two kinds of carp in the United States. People also brought common carp in the 1830s and 1880s. Common carp have since spread throughout US waters. Black carp and grass carp are two other kinds. People brought them to the United States around the same time as bigheaded carp. Black carp have spread much less. And people continue to use grass carp to control water weeds.

end of the Mississippi River. Carp were also spreading through the Illinois, Ohio, and Missouri Rivers. People had found the fish in many US states.

BREEDING POPULATIONS OF BIGHEADED CARP

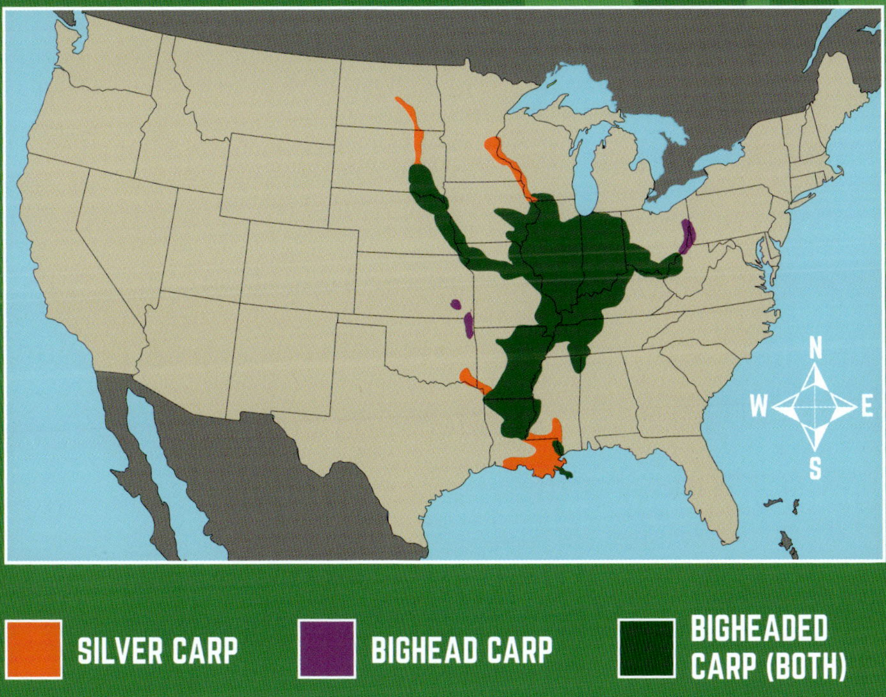

LEAPING FISH

Silver carp tend to feed near the surface of rivers and lakes. Sometimes they become scared. When that happens, the fish often leap out of the water. Silver carp can jump up to 10 feet (3.0 m) into the air. Leaving the water helps them escape predators.

Carp can also become scared by loud sounds. For this reason, they can create problems for boaters. Underwater, the sound of a boat's motor is loud. It can cause the fish to leap into the air. Silver carp sometimes land in boats. They have even hit people. Some people have been hurt by leaping carp. That's because silver carp can weigh up to 60 pounds (27 kg). Such heavy fish can cause serious harm.

People often find the scene of leaping silver carp incredible to see.

CHAPTER 3

OUT OF BALANCE

No species of carp is native to the United States. Non-native species are life-forms that people have brought to a new area. Some non-native species do not cause huge changes. Invasive species are ones that do. Bigheaded carp are invasive. They change food webs in rivers and lakes. These food webs become

Like other fish, young carp are known as juveniles.

unbalanced. Or they may have different numbers of organisms that aren't stable.

In the United States, a few animals eat young bigheaded carp. These predators include eagles and pelicans. However, bigheaded carp produce many young carp

HOW BIG ARE BIGHEADED CARP?

SILVER CARP
UP TO 3.2 FEET (1.0 M) LONG,
60 POUNDS (27 KG)

BIGHEAD CARP
UP TO 5 FEET (1.5 M) LONG,
100 POUNDS (45 KG)

AVERAGE 11-YEAR-OLD
4.7 FEET (1.4 M) TALL,
78 POUNDS (35 KG)

at a time. Young carp grow quickly, too. They tend to reach their adult size after two to four years. The fish can become several feet long. They are too large for most native fish to eat. Adult carp are also too heavy for most birds to eat. As a result, adult bigheaded carp have few natural predators in the United States. Therefore, carp **populations** often grow rapidly.

Bigheaded carp also eat large amounts of food. They can consume up to 10 percent of their weight in food every day. The carp can take away food from native fish. With less food, those fish populations struggle to grow.

Bigheaded carp have changed many river **habitats** in the United States. The Mississippi River was one of the first. This river contains hundreds of species of freshwater animals. But bigheaded carp have taken over large areas of the river.

ALLIGATOR GAR

The alligator gar is a large fish. It lives in the Mississippi River. It is named for its long snout and sharp teeth. Alligator gar are one of the few native predators that hunt adult carp. Bigheaded carp may be large. But alligator gar make them look small. The largest gar on record weighed 327 pounds (148 kg). Some scientists believe gar could help slow the spread of bigheaded carp.

Alligator gar is a popular catch. Some scientists want fishers to catch fewer gar because gar eat carp.

They have crowded out many of these animals. They have caused native fish populations to drop.

CHAPTER 4

TRACKING CARP

People have found bigheaded carp in many parts of the United States. But finding one fish does not mean the species is reproducing there. Carp do not always start breeding in new areas. Sometimes, when carp die, no new carp take their place. For this reason, scientists are most interested in the areas

In the 2010s, silver carp were discovered in several states for the first time. These states were Minnesota, North Dakota, Ohio, and Oklahoma.

where bigheaded carp are breeding. Carp may be in those places to stay.

Bigheaded carp live in groups with other carp. These schools tend to breed together. So, scientists often track carp to find breeding areas. To find the schools, scientists catch a single carp. They place a tracking device on the fish. Then scientists put the fish back into the water. The fish swims back to its school.

This method helps show where bigheaded carp have spread. Then, people can use net systems to remove adult carp. These net systems move slowly through the water. They capture carp as they go. This method does not solve the problem.

In 2018, workers removed more than 240,000 pounds (109,000 kg) of bigheaded carp from a lake in Missouri.

But removing adult carp can slow the population's growth.

Tracking carp helps for other reasons, too. For instance, scientists care about where bigheaded carp have not spread. Using this knowledge, people can try to prevent carp from reaching those areas. For example, the Great Lakes were free from bigheaded carp breeding grounds as of 2020. But people had found carp in nearby rivers.

The Great Lakes hold vast amounts of fresh water. They are home to a huge variety of native plants and animals.

ELECTRIC FIELDS

Only one waterway connects the Mississippi River basin and the Great Lakes. It runs through Chicago, Illinois. Bigheaded carp were most likely to reach the Great Lakes through that waterway. During the 2000s, government officials took action. Workers built several kinds of barriers in the waterway. One kind was electric. It created a field of electricity in the water. Carp are much less likely to pass through this field.

Workers lower an electric barrier into the waterway that connects the Mississippi River basin and Lake Michigan.

Tens of millions of people also rely on the lakes. The lakes provide food, drinking water, and jobs. Scientists worry that bigheaded carp could crowd out the Great Lakes' native fish. That change could cause many problems. For example, people pay to fish in the Great Lakes.

Sport fishers spend billions of dollars every year. That money helps pay people to take care of the food webs in the Great Lakes and Mississippi River.

Scientists believe dams could help keep bigheaded carp out of the Great Lakes and upper regions of the Mississippi River. Carp have a difficult time getting through dams. However, dams have navigation locks. The locks open to let boats pass through. So, scientists are finding ways to stop carp when locks are open. One method uses underwater speakers. Bigheaded carp avoid loud sounds. And some lab studies suggest native fish are less likely to do so.

Water above underwater speakers becomes disturbed when the speakers are turned on.

The speakers act as an invisible fence. Carp are not likely to cross it.

Even so, bigheaded carp are in the United States to stay. That's why the US government created a group to study bigheaded carp. This group supports government scientists. That way, scientists can continue their work in controlling the spread of bigheaded carp.

FOCUS ON
BIGHEADED CARP

Write your answers on a separate piece of paper.

1. Write a letter to a friend describing the main ideas from Chapter 3.
2. Would you rather study bigheaded carp or try to capture them? Why?
3. Where were bigheaded carp first brought to in the United States?
    A. Arkansas
    B. Illinois
    C. Mississippi
4. What would happen to native fish if bigheaded carp spread throughout the Great Lakes?
    A. The number of native fish in the Great Lakes would likely drop.
    B. The number of native fish in the Great Lakes would likely grow.
    C. The number of native fish in the Great Lakes would likely not change.

Answer key on page 32.

GLOSSARY

algae
Tiny plant-like organisms that produce oxygen.

basin
An area of land containing rivers and smaller bodies of water that drain into a larger body of water.

filter feeders
Animals that eat by letting water into their bodies, then capturing and swallowing tiny organisms as food.

food webs
The feeding relationships among different living things.

habitats
The types of places where plants or animals grow or live.

invasive species
Plants or animals that people bring to a new place and that harm people, native plants, or native animals.

native
Living or growing naturally in a particular region.

populations
Groups of animals living in particular areas.

species
A group of animals or plants that are alike and can breed with one another.

TO LEARN MORE

BOOKS

Amstutz, Lisa J. *Bringing Back Our Freshwater Lakes*. Minneapolis: Abdo Publishing, 2018.

Amstutz, Lisa J. *Invasive Species*. Minneapolis: Abdo Publishing, 2018.

Wilcox, Merrie-Ellen. *Nature Out of Balance: How Invasive Species Are Changing the Planet*. Custer, WA: Orca Book Publishers, 2021.

NOTE TO EDUCATORS

Visit **www.focusreaders.com** to find lesson plans, activities, links, and other resources related to this title.

INDEX

Answer Key: 1. Answers will vary; **2.** Answers will vary; **3.** A; **4.** A